Connemara Marble

First published 2014 by The O'Brien Press Ltd,
12 Terenure Road East, Rathgar, Dublin 6, Ireland.
Tel: +353 1 4923333; Fax: +353 1 4922777
E-mail: books@obrien.ie
Website: www.obrien.ie

ISBN: 978-1-84717-597-7

1 2 3 4 5 6 7 8 9 10
14 15 16 17

Designed, edited and published in Ireland.
The paper in this book is produced using pulp from
managed forests.

PHOTO CREDITS

Front cover photographs: scenic: Stephen Walsh; author photograph: Lorcan Brerton, Lorcan Brerton Photography. Back
cover photograph: Mike Bunn.
Internal photographs: Courtesy of: Stephen Walsh: p5, 8, 17, 22, 28-29, 39, 50-51,
56, 57, 69, 73 77, 78, 81, 84, 89, 90-91, 101, 102, 103, 107, 108, 118, 121, 124.
The Balfour Collection, NUI Galway: p10, 12-13, 18-19, 25, 27, 30, 32, 53, 54, 55, 82. Mike Bunn:
p14, 21, 70, 71, 79. The National Museum of Ireland: p35 top right, 95. Ballynahinch Castle Hotel: p35
bottom left and right. Barbara Walsh: p36, 105. The University Club W 54th St New York: p42, 49.
The Chicago Cultural Center: p43. The Bronx Community College: p46-47. Columbia University: p48.
The Clifden Museum: p63. Kevin Joyce: p87. JC Walsh & Sons Ltd: p92, 111. Gonne Wilde: p94.
The John F. Kennedy Presidential Library & Museum (Photo credits: Joel Benjamin): p97.
The Galway Mayo Institute of Technology: p98. Antolini and Co., Verona Italy: p112, 113, 114-5. From *Hall's Ireland*:
p58, 59. From *The Graphic* (1903): p60. From *Memories Wise and Otherwise* by Sir Henry Robinson p64, 65, 66, 67.

Photographs copyright © QVC: p117, 119, 120. Author photograph p122 Lorcan Brerton,

Lorcan Brerton Photography.

TEXT CREDITS

Connemara Marble verses by Gráinne O'Malley

Connemara Marble

IRELAND'S NATIONAL GEM

STEPHEN WALSH

THE O'BRIEN PRESS
DUBLIN

Acknowledgements

My thanks go to my work colleagues who have shared my passion for Connemara marble over a lifetime; and to my father, John, for his vision and wisdom.

A special thanks to Marty Burke and his family in Lissoughter for their endless welcome and support over the years.

I have received invaluable information for this project from my mother, Barbara Walsh, my son Michael Walsh, the Geological Survey of Ireland, the Company of Goldsmiths and Assay Office in Dublin, Kevin Joyce of Recess, Connemara, Dr Patrick N. Wyse Jackson of Trinity College Dublin, Jennifer Gough and Mary Cahill of the National Museum of Ireland, and Mary Vaughan of the Station House Museum in Clifden.

For the visual aspect of this book I am grateful to Marie Boran and Frank Fahey at NUI Galway for permitting the reproduction of Robert Welch's evocative photographs from the Balfour Collection. I am also indebted to Mike Bunn for his Connemara images. For technical photographic assistance, my thanks to Eamonn Cunningham and Sean Walsh, and special thanks to Daniel Walsh for photographic expertise.

In tracing the journey of Connemara marble to the United States, I am most grateful to the University Club in New York, the Gould Memorial Library at the Bronx Community College, the Low Library of Columbia University, and the Chicago Cultural Institute; it was wonderful to see the marble in situ.

In the US also, I thank QVC and, in particular, Jane Treacy, for bringing Connemara marble to a wider audience.

I thank my wife, Marita, for being there always, cheering me on, and guiding this project; and my daughter, Elenor, and sons, Michael and Daniel, for being supportive and bringing that special wit and wisdom to bear when most needed.

Finally, sincere thanks to Michael and Ivan O'Brien of The O'Brien Press, and to editor, Helen Carr, and designer, Emma Byrne – all of whom have been a delight to work with.

'what in me is dark
Illumine, what is low raise and support;'

John Milton

Paradise Lost. Book I

(Inscription above Connemara marble pillars in the

Gould Memorial Library, New York)

CONTENTS

Foreword

It was a glorious day in Ireland. Blue skies, warm sunshine and green hills welcomed us to Connemara, one of the most stunning regions of Ireland. I was there with a crew from QVC in the early 1990s, and we were going to shoot with a new guest who was bringing Connemara marble jewellery and gifts to QVC. We arrived at the site, to hear a booming voice call, 'One hundred thousand welcomes to Connemara!' It was Stephen Walsh himself! He had the widest smile and a twinkle in his eye, and he wove magical stories about the fairies bringing the forty shades of green to the marble, and how the mine was something out of an Irish fairytale. Our friendship was cemented later that evening when I sang 'Forty Shades of Green' in the lounge at our hotel and Stephen leaned over to tell me to keep my day job!

Now, over twenty years later, I am proud to call Stephen one of my dearest friends, even though we are often separated by the Atlantic Ocean. We share a love of Ireland and have the honour of sharing that passion with the wonderful viewers of QVC. I treasure our time spent on air with the beautiful Connemara jewellery. (Did you know it is 900 million years old? Stephen will quiz you!)

As you read these pages know that they are written in the true Irish spirit of love, loyalty and friendship. No one embodies that more than my friend Stephen Walsh, who brings the luck of the Irish to you!

Sláinte!!

Jane Treacy

QVC

9

Introduction

There are whispers of history in the wind.
Mist smelled in the air, in the crumbling
dense black turf,
And precious stone buried in the mountains.

IRELAND'S NATIONAL GEM

Connemara is the windswept and heart-stoppingly beautiful region that lies in the westernmost part of Ireland, facing out over the Atlantic Ocean. Beneath this sharp, rugged landscape lies the green marble, a rare, unique and iconic stone; a real piece of Ireland carved out from this special place. Throughout the ages it has been highly prized in Ireland and beyond, and has been witness to the fascinating and often strange history of this enchanting region.

I have the privilege of owning seven acres in Connemara, and this has introduced me to a people and a lifestyle that I would otherwise have missed. A Dubliner by birth, I have spent my working life in the family business, J.C. Walsh & Sons Ltd. It was luck that brought me to Lissoughter, and good fortune that I have been able to grow a business, and see the rise of Connemara marble to its status as an iconic Irish brand.

All of this has led me on a marvellous journey, and I am thrilled to be able to share some of this excitement with you. Let me be your guide in taking you through the history and lore of this unique marble, and introducing you to people, events and places it has touched. Although the marble has travelled far and wide over the centuries, its story has never been told in one place before. Come with me now as we discover where the tale of Connemara marble began.

Beneath this sharp, rugged landscape

lies the green marble, a rare, unique

and iconic stone; a real piece of Ireland

carved out from this special place.

Connemara Marble – Geology

Marble twisted and shaped before
time began. Form and beauty
buried in simple stone.

Connemara marble was formed in the uncharted history of geological time; in nature's slow upheavals in the long past – unseen by eye, unheard by ear – before any timepiece measured time as we know it. In that primordial upheaval the Earth churned and boiled and cooled; oceans altered and landmasses shifted over an expanse of time so vast that our era, the age of man, is but a grain of sand to its deserts, a mere drop in its primal ocean.

The word 'marble' is derived from the Greek word '*marmaros*', meaning 'shining stone'. In its geological sense, the term applies to sedimentary rocks – rock formed from the settlement of silts and then metamorphosed or changed by heat and pressure into marble.

It is in the pre-Cambrian era, 900 million years ago, that the story of Conne-mara marble begins. The marble occurs in what is known as the Appian group of rocks (a sub-group within the pre-Cambrian 'Dalradian Supergroup', which runs in a northeast to southwest direction from Scotland into Ireland, in the counties of Donegal, Galway and Mayo, all on the western coast). The minerals that combine to give Connemara marble its unique colours – sea greens and deep greens, swirling wave greys, bronzes, coral pinks and flashes of gold – began forming at this time; they were present in the now vanished Iapetus Ocean (a precursor to the Atlantic Ocean). As time passed these minerals became deposits of limestone and quartzite, around 600 million years ago.

There followed the Grampian or 'mountain building' era (580–475 million years ago) in which the sediments were folded and refolded. Intense heat and pressure caused the deposits to change their structure to become marble, while the folding and refolding that occurred led to the unique patterns and textures in the stone.

The diverse colours found in Connemara marble come from the minerals serpentine (deep jewelled green), epidote (the many paler shades of green), micaphlogopite (bronze), calcite (the misty greys) and pyrite, (a touch of gold). Pyrite, or iron pyrite, is an iron sulphide and its metallic lustre and pale yellow hue has earned it the nickname 'fool's gold'. The name 'pyrite' comes from the Greek *'purites'*, the name given in ancient times to materials that would create sparks when struck against steel. Pyrites form crystal-like structures within the marble, and can enhance the colours of the stone. But it is serpentine that gives the marble the darkest, most lustrous and translucent shade of green. The inten-sity and vibrancy of the colours in the marble attracted the early settlers long ago and endowed the stone with value and prestige.

Thus, Connemara marble was formed in nature's crucible – heated and cooled, engulfed by time, and left to sleep. This beautiful green marble occurs only in a thin band running across Connemara just north of the main Galway

A megalithic standing stone

to Clifden road, from Lissoughter in the east to Streamstown in the west. The marble rises to the surface in a handful of locations, and these points became the initial focus of human interest, the markers for future quarries.

And what could be more special than Connemara marble? Its colours, swirls and waves reflect the very topography of the landscape that produced it. Each slab is singular in its pattern, each piece truly one of a kind and distinct as a fingerprint. It is said that Ireland has forty shades of green, and the myriad of greens in Connemara marble make this stone a true treasure – Ireland's national gem.

Thus, Connemara marble was formed in nature's crucible – heated and cooled, engulfed by time, and left to sleep.

Connemara – Early History

There is a quiet corner of my soul that sits apart,
And though the world may tumult elsewhere
This holds peace.

Connemara is a region of County Galway on the west coast of Ireland. It occupies an area of roughly two thousand square kilometres (some eight hundred square miles) and is an area of stunning beauty and breathtaking landscapes.

The Twelve Bens mountain range dominates the region, and the rugged Atlantic coastline is filled with intricate bays and hidden coves. The land is poor – mostly bogland – and there are countless lakes, streams and rivers wending their way through the region. The Irish language is widely spoken by many of the inhabitants, and it is from the native lexicon that

These Mesolithic tribes lived in simple settlements, but they were a thriving, well-developed society.

many of the local place names are derived. In fact, the very word 'Connemara' comes from '*Conmaicne Mara*' – '*mara*' being the Irish word for 'sea' – the name given to the branch of the Conmaicne tribe who lived the by the sea over a thousand years ago.

The region is moist, with Atlantic weather systems bringing a temperate climate and above average rainfall. The poor land of the wild and beautiful landscape can scarcely support a few hardy sheep, but in hillside hollows and sheltered corners away from the harsh Atlantic winds, land has been cultivated, and a patchwork of small fields, known locally as 'gardens', make the most of the thin Connemara soil.

Early settlement

Local historians have traced the settlement of man in Connemara dating back seven thousand years, and possibly as far back as ten thousand years, to the Mesolithic, or middle Stone Age. The early settlers lived in the coastal areas, and evidence found near the present-day town of Clifden, points to a society dominated by fishing, hunting and trading. These Mesolithic tribes lived in simple settlements, but important finds of tombs, standing stones, ancient trackways and cooking sites lead us to believe there was once a thriving, well-developed society. Further archaeological settlements and finds have been discovered dating to the Bronze Age and the Iron Age, including *crannógs* (lake dwellings), ring forts and tombs. It is estimated that sea levels have risen over five metres during the last five thousand years. As many settlements were in coastal areas, it

Opposite: A megalithic standing stone

can be assumed that there are many, as yet undiscovered, sites holding evidence of this early society in the sea depths.

The earliest evidence for the use of Connemara marble dates back to the Neolithic era, over five thousand years ago. Professor Sean P. Ó Ríordáin of University College Dublin carried out a series of excavations at Lough Gur in County Limerick in the 1940s and 50s. His excavations uncovered evidence of prehistoric dwellings and many items associated with the people who lived at that time; items such as pottery, axes and decorative beads were all found. Most of the stone beads were made from siltstone, but a significant number were green serpentine and are reckoned to have come from Connemara. Clearly, early man recognised the beauty of this rare and unique green marble, and it played a role of special importance in decorative and ceremonial use.

There are other finds too. A spectacular polished porcellanite (green stone) axe head, dating from the Neolithic era, was found in the village of Cashel in south Connemara. This axe, one of the finest in the country, was found hidden in a cleft in a rock, face down, and may have been a symbolic offering rather than a misplaced or discarded tool.

CHRISTIANITY

The arrival of St Patrick in 432 brought Christianity to Ireland, and it is told that the saint journeyed through Connemara in 441 on his way to Croagh Patrick in County Mayo. Patrick's journey in Connemara brought him on the 'pilgrim's path' (a place less than ten miles from the marble quarry at Lissoughter). At a height of about 1,200 feet, in a windswept gap known as the pass of *Mám Éan* (from the Irish, 'the pass of the birds'), there is a holy well and place of pilgrimage; an annual event still takes place at this ancient place of worship. This holy well, *Tobair Phádraic* – St Patrick's Well – has an ancient tradition that

The Twelve Bens and Roundstone Bay
from Station Island.

one should walk seven times around the well whilst reciting prayers, and throw a stone onto the enclosure each time. An altar of Connemara marble has been built at the site, and Mass is still celebrated at this remote and desolate location.

The tradition of the Holy Well is particularly strong in Connemara, where wells are dedicated to Saints Colmcille, Macdara, Feicin, Ceannach and Cailin. There are several monastic islands in Connemara of which one of the most interesting is Saint MacDara's Island. This contains a well-preserved oratory, an early graveyard and a holy well.

VIKINGS AND NORMANS

The first recorded attacks on Ireland by Vikings are for the year 795. The raiders attacked several island monasteries off the coast. They made little impact on Connemara, although evidence of their presence has been found in the form of weaponry and armour near current-day Clifden, and on Inisbofin, an island some five miles off the Connemara coast.

The Normans, too, failed to conquer Connemara, but they did shape its history – particularly the de Burgos, who pushed the Gaelic O'Flaherty Clan west into Connemara as they advanced. The O'Flahertys (known as the 'furious Flahertys') dominated Connemara up to the time of the Reformation and built several castles in the area, which can still be seen in the landscape. Hen's Castle in Lough Corrib and Renvyle Castle were among the O'Flaherty settlements. While the Normans never captured Connemara, their influence pervaded, as is attested to by the prevalence of Norman names – Joyce, Burke, Walsh, Barry and Staunton, to mention but a few – which still thrive today throughout the area.

The O'Flaherty lands were confiscated in the seventeenth century, during the Cromwellian Wars, when many Gaelic families lost their landholdings. They continued undeterred, however, and with their characteristic sense of adventure, they turned to new skills, including illegal trading in wool, wine, tobacco and sherry. They conducted a prosperous business in ambergris – which comes from the sperm whale and is used in the making of perfume. This valuable commodity was found floating on the Atlantic waters and had a ready market in Europe and beyond. Shark fishing and the harvesting of kelp brought prosperity too, and a lively commerce blossomed. There are hundreds of hidden coves and jetties along the rugged coast where illicit trades could take place unseen. Many coastal placenames in Connemara still attest to the smuggling that was carried on centuries ago: Brandy Cove, Brandy Harbour, *Duirling an Eabhair* (Ivory Shore) and *Duirling an Chadais* (Cotton Shore).

In the mid-seventeenth century a branch of the Anglo-Norman family Martin, one of the Tribes of Galway, was granted O'Flaherty lands at Ballynahinch, in Connemara. Further west, additional O'Flaherty lands went to the D'Arcy family (who would found the town of Clifden and reside at Clifden Castle).

These two families, the Martins and the D'Arcys, would effectively dominate the social, political and economic life of Connemara for the next two hundred years.

A kitchen midden with limpet shells, Dog's Bay, Connemara.

As many settlements were in coastal areas, it can be assumed that there are many, as yet undiscovered, sites holding evidence of this early society in the sea depths.

Beginnings
Of Quarries

The search for pure, translucent green — the vein of serpentine flowing through the stone. The rarest Connemara marble of all.

It was in the early nineteenth century that Connemara marble was rediscovered and its potential exploited. This development coincided with a surge of building in England. At the time, Britain was at the height of its power and the Empire was thriving. A period of calm and prosperity led to a demand for great buildings with ornate and decorative interiors in Gothic, Renaissance and Classical styles. Traditionally, marble was sourced from Italy and Greece, but transport costs and difficulties with supply were a prohibitive factor.

It made sense to look to Ireland for a source of raw materials since firstly, it was so close, and secondly, it was under British rule and therefore free from trading obstacles. Various tours were made to Ireland on behalf of British rulers to ascertain its economic and natural assets and their value to the crown. It made practical and economic sense to utilise any resources close to home.

31

Previous page: a cottage at Cashel Bay.
Above: A mountain farm, Ballynahinch.

In *Hall's Ireland*, published in 1843, there is mention of a quarry of black marble located on the banks of Lough Corrib, near Galway. The book tells of an English man (whose name and occupation are lost) exploring the country for minerals, and who chanced to discover a stone of fine texture, which on polishing by a mason was pronounced marble of a fine colour. This discovery induced two local brothers to export a cargo to London, where it met with immediate sale among the merchants at a high price. While it is difficult to be precise as to when the quarries in Connemara were cleared and opened, it is fair to say that all available evidence points to them being opened in the early nineteenth century.

The D'Arcy family

The quarry at Streamstown is located about three miles north of Clifden, and the local landlord, John D'Arcy, was responsible for opening it commercially. There was no roadway connecting Clifden to the city of Galway some forty-eight miles away, and access was only possible on horseback until a road was finally opened in the 1830s. Faced with this problem, D'Arcy saw that the only way to promote commerce and trade was to develop a harbour in Clifden. Funding was secured and the building of a new harbour commenced in the early 1820s. There was widespread support among the population for this project as it was felt that future development depended on it. In the summer of 1824, with the quay not yet fully finished, the first cargo of marble set sail from Clifden. It was a cargo from the Streamstown quarry, bound for Liverpool. To celebrate this momentous event every house in the town was illuminated that evening.

As a landlord, D'Arcy offered leases to prospective tenants to work lands, fisheries and quarries. Streamstown was leased to the Hibernian Marble Company for forty-one years at an annual rent of £350, or one sixth of the profits made. However, it seems that only one or two payments were made, and by 1830 the lease was again on offer. It seems that the D'Arcy family themselves took over the running of the quarry for some time, and it was offered for sale in 1850 after the collapse of the family's fortunes. In 1895, it was acquired by Robert Fisher, from New York.

In Hall's Ireland, *published in 1843,*

there is mention of a quarry of black marble

located on the banks of Lough Corrib.

The Martin Family

The other great Connemara family, the Martins, held large estates east of the D'Arcys. Their family home was at Ballynahinch Castle in the heart of Connemara. The discovery of marble in their estates at the townlands of Barnanoraun and Lissoughter prompted the family to open quarries there in the early 1800s.

Humanity Dick Martin

One of the great characters of Connemara presided over the Martin family in the early nineteenth century. Richard Martin MP (1754–1834) was known as 'Humanity Dick'. His nickname derived from his concern for the protection and welfare of animals and the practical application of his philosophy; he kept a prison on an island in Ballynahinch lake where he detained those who were cruel to animals on his estates before there was legislation to prevent such deeds. He succeeded in getting an Act for the protection of animals through parliament – the first of its kind anywhere, and it was from this that the RSPCA grew. This eccentric gentleman was a larger-than-life character, lampooned in poetry and cartoons of the time, and known widely as the 'King of Connemara'.

Like many of the landlords at that time, Humanity Dick's finances were in poor shape, and work was carried out to exploit resources on the estates. In this context, Humanity Dick hoped that the marble quarries would be an asset and help to improve his finances – he is the first person credited with having Connemara marble polished. In fact, one the earliest examples of the commer-

Opposite: a marble-topped table, marble fireplace and detail from a marble fireplace from Ballynahinch Castle.

cial production of Connemara marble that can be seen is a fine pair of matching tables (currently on display in the Dublin Decorative Arts and History Museum). These tables are said to have been made at Ballynahinch Castle, and were presented by Humanity Dick Martin to Ann O'Hara of Raheen, Galway on her marriage to James Bourke in 1817. His son, Thomas, had two oval topped tables made for Ballynahinch Castle, and Humanity Dick himself is said to have presented a Connemara Marble fireplace to the King.

The Martins had successfully built a road from their estate to Galway, but it was unsuitable for the transportation of heavy goods. Accordingly, as the marble was lifted from their quarries, it was taken over a steep road of some six miles to Cloonisle Pier for transportation to Dublin and beyond, for cutting and polishing.

The transportation problem would eventually be alleviated significantly by the advent of the railway. The Midland and Great Western Railway Company developed a rail route linking Clifden to Galway with a grant of £264,000. Work commenced in 1891 and the first train reached Clifden in 1895. There was even a station at Ballynahinch. The route brought

freight and visitors through one of the most untouched landscapes in the world.

In 1826 Humanity Dick lost his parliamentary seat and fled to Boulogne in France, as he could no longer enjoy parliamentary immunity to arrest for debts. He died in France in 1834. Twelve years later Ireland was struck by the great famine, Humanity Dick's son Thomas died of famine fever, the estate became bankrupt and the Martins no longer presided over the estates at Ballynahinch.

No story about Connemara is complete without mentioning Kylemore Abbey, the landmark gothic building situated to take maximum advantage of the surrounding mountains and lakes. Kylemore was built by an Englishman, Mitchell Henry. He and his wife, Margaret, spent their honeymoon in Connemara in 1852, and were very taken with the beauty of the landscape. In the 1860s, Mitchell Henry inherited a substantial amount of money; he returned to Connemara and purchased the lands that are now the Kylemore Estate. He set about creating a home in his beloved landscape in the heart of Connemara, and thus the building of Kylemore Castle commenced in the late 1860s. Mitchell and Margaret came to live there, but sadly, Margaret died in1875 at the age of fifty, and after her death, Mitchell had a church built on the estate in her memory. The church is built in the gothic style, and despite its relatively-small size, it has the visual impact of a cathedral in its form and style. It is said to be reminiscent of Bristol Cathedral from the inside, and Norwich Cathedral from the outside. The interior is decorated with marble columns using a number

Kylemore Abbey (opposite), the landmark gothic building situated to take maximum advantage of the surrounding mountains and lakes.

of different marbles and, of course, Connemara marble is included. Kylemore passed into the ownership of the Benedictine nuns in 1920. The nuns established a successful boarding school for girls there, but dwindling numbers in the community meant that the boarding school eventually closed in 2010. The Benedictine nuns still own Kylemore, and it is now a very successful tourist attraction. The castle and church are both open to the public, and there is a Victorian walled garden in the grounds, which has been restored in recent times. It is indeed a remarkable destination. An oasis of romance and gothic splendour set against the rugged beauty of Connemara, it is commensurable with any tourist destination worldwide.

Connemara marble attracted a good deal of attention in the nineteenth century. Sir Charles Lewis Metzler Von Gieske toured Connemara and submitted a report of his findings in 1826. These were subsequently published in 1834 in the Dublin University magazine. Here is an interesting extract from that report:

The day after my arrival in Galway I went on an excursion to Connemara. In the evening I arrived at Ballynahinch, the residence of Thomas Martin Esq., who received me with his usual kindness and on the following day accompanied me to what is called the Green Marble Quarry, but is rather a quarry of precious serpentine, belonging to his estate. It is situated in a valley extending from the north to the west peak of Lettery – as far as the middle of 'the Twelve Pins', a series of very pointed primitive mountains. I found, on following up the river, or rather torrent, traces of serpentine, at a distance of a mile from the quarry. The river, which takes a serpentine direction, has disclosed to the eye extensive strata of the most beautiful granular marble of a pearl white colour, mixed with rose red, blood red, yellowish red and blue-grey alternating with green stone. The serpentine quarry, where Mr Martin keeps 150 to 170 labourers employed in blasting, cutting, and sawing the immense blocks, is of an extraordinary extent, and seems to be inexhaustible. The serpentine, similar to the serpentine of *antico* of Italy, is mixed with

Cloonisle Pier

steatite, fine granular limestone and stripes, and occurs in blocks, sometimes at a length of twelve or thirteen feet, and three or four tons in weight. It is impossible to describe the immense varieties of delineations, shades and colours of this beautiful stone, which attracts the eye of the beholder. The serpent-like veins of some excite particular admiration, others are coloured in spiral forms; others are dotted and spotted with different shades of green, grey and yellow. Solid masses of an enormous size may be raised.

Mr Martin has already quarried out an immense quantity, part of which is cut in slabs for tables, which are ready for sale. He has made a road from the quarry to the port, a distance of six miles, but it would require a rail-road for large blocks.

In John Jordan's *Topographical Dictionary of Great Britain and Ireland* of 1833, mention is made of activity in the Connemara marble quarries. In 1851 at the Great Exhibition in London, several specimens of Connemara marble were on show, mainly in the form of table tops and fireplaces. The famous geologist George Henry Kinahan, (1829–1908) toured Ireland in 1878 and reported his findings about the quarry at Lissoughter as follows:

Immediately east of the village is the serpentine marble quarry belonging to Messrs. Sibthorpe of Dublin who raise and square the blocks here, and cart them thence to Cloonisle pier, where they are shipped for their manufactory in Great Brunswick Street, Dublin. The rock is known in the market as the 'Lissoughter green marble', and we are informed by Messrs. Sibthorpe that 'its average value in the rough in Dublin is about sixteen shillings per foot cube for fair-sized blocks'.

In a further report in 1889, Kinahan talks about 'brooches and other articles of vertu being made from Connemara marble'.

An edition of the *Irish Naturalist Journal*, Vol. 4, No. 2 (Feb 1895) contains on article with the intriguing title 'The Geologist at the Luncheon Table', in which

Professor Grenville A. J. Cole describes some marble tables that had been gifted to the refreshment rooms at Trinity College Dublin:

> Naturally, the polished tops have been made of Irish marbles under the care of Mr E. S. Glanville, of Lower Erne Street, Dublin. A scientific committee of selection visited the works and the stones were chosen as if they were to ornament a museum. While literary and antiquarian visitors cannot fail to appreciate the artistic beauty of the slabs, naturalists will find many points of interest, even in their minuter details.
>
> Two of the slabs are from the Lissoughter quarry in Connemara, and show the unique serpentinous marble in all the perfection of its green and grey streaks and folding. The highly metamorphosed character of the rock is at once apparent, and in one table the contortion of the bands can be traced, while in the other a more parallel structure has been set up by the continued deformation of the mass.

Another edition of the *Irish Naturalist* carries a description of the quarry in Lissoughter:

> The celebrated Connemara marble is now being worked in only one quarry by about a dozen men, employed by an American who is at present executing an order for twenty pillars, each to consist of five blocks measuring about four feet by three in diameter. To gauge the quality of the stone, a long section has been made in the side of the quarry by a wire saw. This section well displays the contorted green bands with grey layers at either side. The so-called marble is a serpentine formed by the alteration of olivine introduced in the crystalline bands of primitive limestones by igneous action. Similar results occur in limestones around Mount Vesuvius in Italy.

In 1893, Connemara marble was to be found on display as far away as Chicago at the World's Fair. The fair was a social and cultural event, and influenced

architecture, the arts and American industrial optimism at the time. Designs of the day favoured the Neo-Classical, with its symmetry, balance and splendour. The famous glass-making firm, Tiffany and Company, created an exhibit to 'put before the eyes of visitors various objects from different departments, to illustrate the scope and business done by the firm, which embraces all the forms of ecclesiastical and domestic embellishment.' Within this spectacular exhibit, Tiffany and Co. displayed 'sacred vessels, candlesticks made with gilt and Connemara marble, filigree work and precious stones, a marble pulpit ...'

Four years later, the Chicago Public Library (now the Chicago Cultural Center) was completed – a spectacular showcase building of its day. Designed by the Boston architects Shepley, Rutan and Coolidge, it was the apotheosis of the aspirations that had been manifested at the World Fair. The architects chose the most sumptuous materials, imported the most rare marbles and decorative items, and used the skills of expert workers to create an architectural statement. The lobby is a dazzling, light-filled space covering three stories, and featuring Cosmati work – mosaics of glass, gold leaf, mother of pearl and precious stones. The balustrade railings of the staircase are divided into panels adorned with this work, and each has an eight-inch disk of ornate dark and pale Connemara marble.

Opposite: The lobby of the University Club, New York.
Below: Detail from the Chicago Cultural Center.

Thus, one can trace a great deal of activity associated with the quarries in the nineteenth century as Connemara marble was discovered by the wider world. The increasing demand for the marble provided employment in the Connemara region, and the stone was used both at home and abroad in many illustrious building projects. Examples of its use in Ireland at this time are the museum of geology in Trinity College Dublin (1857), and St Mary's Church, Pope's Quay in Cork (1872). Six slender Connemara marble columns adorn the entrance hall of Farmleigh House, located in the Phoenix Park in Dublin. This fine house was built in 1884 by the Guinness family and is now in State ownership and used to accommodate visiting dignitaries and guests of the Nation as well as some Government meetings. In England, Connemara marble is featured in a great many churches built or restored during the nineteenth century, including the presbytery of Great St Mary's Church (1866), Gloucester Cathedral (1870–1872), Stetchworth Church (1876), the presbytery of Worcester Cathedral (1877), St. John's Church, Waterbeach, near Cambridge (1878), Truro Cathedral (1886), Peterborough Cathedral (1892), Bristol Cathedral (1895). It was also used in the Ashmolean Museum (1840s) and the Pitt Rivers Museum (1886), both of which are in Oxford, in pillars at St John's College in Cambridge, and in Kensington Palace, now home to the Duke and Duchess of Cambridge.

It is in the United States, however, that many outstanding examples of Connemara marble building work are to be found. By the close of the nineteenth century, the United States was a wealthy, modernising and technologically-advancing force, and a leading world power. This new confidence found its expression in cultural, economic and artistic spheres, and was witnessed architecturally by a wave of city building across the States. Much of the architecture was expressed in Neo-Classicism, drawing parallels between the young, vibrant nation on the one hand, and the Old World – Ancient Greece and Imperial Rome – on the other. Notions of democracy, law, justice and liberty found architectural expression in the grandeur of classical forms. The architectural firm of McKim, Mead and White flourished at this time, and in architectural terms, the period

1880–1917 is referred to as the American Renaissance.

Two buildings in New York built during this American Renaissance contain what must surely be the finest examples of Connemara marble anywhere. The first is the University Club at West 54th Street, which features an entrance lobby completely clad in this marble from the Lissoughter quarry. It is Romanesque in style, recalling the peristyle to a traditional Roman villa. The fifteen-foot solid marble columns are made of single pieces of Connemara marble – an extraordinary feat, given the nature of the marble and the difficulties of transportation at the time – and the veining and colours are extraordinarily striking and majestic in their beauty. The building, erected in 1899, was designed by the famous architectural firm of McKim, Mead and White and is without doubt one of New York's grandest buildings.

The second building is the Gould Memorial Library, designed by Stanford White of McKim, Mead and White, for New York University's new rural campus and opened in 1900. White, who collaborated closely with the University's chancellor, Henry Mitchell McCracken, used the Pantheon in Rome as his inspiration. The early design called for sham marble columns in the reading room, but in a letter dated 3 September 1896, McCracken states 'I would have esthetic pangs every time I saw them.' Eight days later, Stanford White proposed the use of Connemara marble columns:

I find that columns in the Library Dome can be made of Connemara Irish Green Marble for forty-one thousand dollars …

… It is the most beautiful green marble in the world, and it would be a great thing to use it …

Interestingly, there was a bit of inter-university rivalry going on in the margins of this tale. At the same time, a new campus was being designed for Columbia University by Stanford White's partner, Charles McKim. Columbia had sought eighteen Connemara marble pillars for its own library in the new campus in Morningside Heights in Upper Manhattan, but was only able to obtain two – probably due to difficulty in quarrying pieces of the diameter required. Columbia's two extraordinary Connemara marble columns were, however, placed in the most prominent position within the Low Library (named after their president, Seth Low) forming a screen that is visible as soon as one enters the building. The critic, Charles H. Caffin, remarked of them:

> No description can give an adequate idea of their stateliness, the exquisite mystery of graded greens and grays and black, their tempestuous streakings and tender veining, and the perfect texture of their polished surface. The most heedless visitor cannot pass them unadmired, the connoisseur will be enthusiastic.

Meanwhile, both Stanford White and Chancellor McCracken could be pleased indeed by the gracious columns obtained for New York University's new library. Sixteen thirty-foot columns of polished, deep-green Connemara marble from the Lissoughter quarry support a circular entablature and limestone balcony on which classical statues are stationed. This is topped by the sixty-foot span of a hemispherical gilded dome, at the centre of which a huge oculus opened (now, sadly, replaced by floodlighting) providing the only sunlight in the rotunda. The library itself is a metaphor for learning and intellectual illumination – light reaches into darkness – and the marble columns are the support for the entire sentiment expressed, not only architecturally in the building, but also in the words from Milton's *Paradise Lost* around the architrave above their capitals:

What is dark in me, Illumine, what is low, raise and support.

The importance of the Gould Library has been recognised by its inclusion in the US National Register of Historic Landmarks.

The journey of the Connemara marble across the Atlantic to New York to become part of the fabric of these peerless buildings – both outstanding examples of the use of this ancient stone, unmatched anywhere in the world – is indeed a remarkable one.

Opposite: Construction workers with a marble column in the Low Library, New York. Below: the lobby of the University Club, New York, c.1900.

It was in the early nineteenth century that

Connemara marble was rediscovered

and its potential exploited.

Development Of Connemara Marble Jewellery And Souvenirs

*Nine hundred million years the marble lay
beneath the mountains of Connemara.
Until craftsmen shaped it for those of us
who have a passion for beauty.
For the past.
For the wild and lonely soul of Ireland.*

The word 'souvenir' comes from the French, meaning 'keepsake' or 'remembrance'. The practice of purchasing a keepsake or small memento of a journey can be traced back to Greek or Roman times when small stone pottery figurines were produced by local craftspeople for sale to visiting travellers and pilgrims.

A Connemara long car.

Building the railway at Recess. Opposite below: wool spinning in Connemara.

Citizens of the Roman Empire considered travel to important parts of the Empire to be an integral part of their education. As Christianity became more widespread, the idea of pilgrimage became increasingly popular. Journeys were undertaken to holy sites and churches, to visit shrines and to see the relics of saints. Of course, the pilgrim wanted to mark the completion of the journey, and it became common for returning pilgrims to wear a lead badge indicating where he or she had been. Accordingly, a bust souvenir industry developed around the cathedrals and shrines, and typical examples of mementos would be a palm from the Holy Land, a scallop shell from Santiago de Compostela in Spain, or a set of keys from Rome. By the seventeenth century wealthy young men were undertaking what was known as the 'Grand Tour' of Europe, but it was not until the industrial revolution and the advent of the railways that leisure travel began to be enjoyed by the middle classes.

During the Victorian era, the idea of travel as leisure became increasingly popular. Given its proximity to England, the Irish countryside made an interesting destination for English and European visitors. The brilliant Scottish engineer, Alexander Nimmo, developed a carriage road in Connemara, linking Galway and Clifden, in the nineteenth century. This allowed for a twice-daily coach service bring mail, provisions and tourists. The building of the railway from Galway to Clifden started in 1891 and the first passenger trains commenced in 1895. In order to promote tourism, the railway company built a hotel at Recess, twelve miles from Clifden.

In the 1843 book *Ireland, its Scenery and Character*, the authors, Samuel Carter and Anna Maria Hall, describe their travels and provide valuable insights into Irish places, lifestyles and customs. Fortunately for us, County Galway, and more specifically, Connemara, are included in the Halls' itinerary. They were enchanted with the natural beauty and wildness of Connemara, describing it as 'the Highlands of Ireland'. They also mention the marbles as an important resource of some value. They note:

The great intermixture of serpentine and talc in all the rocks of this wild region, distinguish them remarkably from those of the rest of Ireland … Precious serpentine, of various shades of green and yellow, often mottled and striped, is intermixed with the white and rose-coloured limestones; and a very beautiful marble is thus produced, precisely the same structure and appearance as the *verde antico* of Italy, and undoubtedly the richest and finest ornamental stone yet found in these kingdoms. The most beautiful varieties occur at Ballynahinch and Clifden in Connemara, where extensive quarries are, unhappily, but partially worked. It is much to be regretted that this beautiful marble is so little known. There are decided indications of its existence in other parts of the same district.

Another English visitor to Ireland, Theresa Cornwallis West, wrote of her travel to Ireland in the book *A Summer Visit to Ireland, 1846*. In it, she describes visiting a shop where she admired numerous souvenirs including a gold-mounted walking stick carved with wreaths of shamrocks and priced at twelve guineas, a model of an ancient Irish harp, bracelets and brooches set with Irish gems, as well as studs, pins and rosaries. By the 1850s, Connemara marble jewellery was widely advertised, and it became the inspiration for Scottish 'pebble' jewellery where stones were cut into panels set flat in engraved silver.

An interesting glimpse of the hardship caused by the Great Famine, of the 1840s can be found in a pamphlet, published by J. Wilson Browne in 1847 appealing to the British public for the promotion of industry in Ireland. It describes the abject poverty found in the town of Clifden and how the failure of the potato crop led to all business being destroyed; how fishing tackle was pawned to buy food, and how the people no longer had capital to buy wool to knit garments for sale. In relation to Connemara marble it states:

> The Connemara marble is found in great abundance, and although it makes beautiful ornaments, which are very saleable, there is only one man in the neighbourhood who works it. He has not the capital to buy a wheel, and cuts and polishes whatever is manufactured, by the slow process of hand labour.

A further book about tourism in Connemara and the surrounding regions was published in 1871. Entitled *Connemara and the West Coast of Ireland: How to see them for six guineas* by John Bradbury, it suggests an upsurge in tourism for those travelling on a budget, and is probably paralleled today by tourist guide books with 'on a shoestring' or '$5 a day' in the title. Bradbury's book covers in great detail how to tour these remote parts of Ireland, with careful explanation of trains and coach fares, distances covered and suitable lodgings. He writes a charming account of his stay in Clifden, and gives an interesting description of the local trade in marble items:

> Near to the hotel [in Clifden] on the opposite side of the street is the shop of Alexander McDonald, a native of Clifden who works very pretty brooches and other articles out of the green serpentine marble that is found in the neighbourhood. His charges are reasonable, and by visiting his establishment, pretty souvenirs of your journey can be purchased.

Connemara marble souvenirs: Donkey (1960s), harp (1910), shamrocks (Victorian era).

A tour guide to Clifden published in 1877 describes a visit to the region as follows,

Clifden can be reached, as intimated, from Roundstone, or, returning by Lough Ballynahinch, the more picturesque route can be resumed. After leaving the Lough, a road to the right leads to the green marble quarries, where a finely-streaked calcareous serpentine mottled stone with various shades of green and white is found … It is especially adapted for ornamental columns and mantles, and is worked onto other forms, brooches, crosses, etc., as souvenirs of a Connemara excursion.

In 1849, on a visit to Dublin, Queen Victoria was presented with a copy of a traditional Irish brooch known as the 'Cavan Brooch', which she immediately wore and which became known as the 'Queen's Brooch'. Queen Victoria also wore a Claddagh ring – the traditional Irish ring with two hands clasping a heart and crown – giving further endorsement to the popularity of Irish jewellery.

Connemara marble featured in jewellery from the outset among the typical Irish souvenirs. In fact, alongside the *Galway Express* coverage of the Royal visit by King Edward VII in 1903, there is an advertisement placed by Faller's Jewellers of Galway announcing that they have a wide range of Connemara marble jewellery on sale.

The green marble made an appearance in another context too. Soldiers of the Great War (1914–1918) sometimes carried a luck charm. In particular, Irish-born soldiers would carry a four-leaf clover made from Connemara marble; some of these luck charms can now be seen in the Imperial War Museum at

Burlington House in London. Interestingly, the name for young clover in Irish is '*seamairóg*', anglicised as 'shamrock'. The shamrock and the clover are of the same family, and the three-leaf shamrock was a common motif in many Irish souvenir items. Thus, the shamrock has a double association, firstly with good luck and, secondly, with St Patrick, who was said to have used it to explain the Christian mystery of the Holy Trinity.

These were the origins of a market for unique Irish souvenirs and gifts using Irish materials and designs. Shamrocks, harps, high crosses, sorrowing maidens (symbolising Erin), wolfhounds – these were the popular souvenir emblems that featured in the early designs. Connemara marble quickly became part of the tradition of Irish giftware alongside knitwear, crystal and tweed; it found its place easily among the plethora of indigenous industries and crafts that grew up around the business. It was an ideal material for use in souvenirs not only by virtue of its innate beauty, but also because of its green shades, which evoked the landscape it commemorated in the souvenirs themselves.

The Royal Visit.

The Royal
Visit

*Light never so green, so blue, falling
over the mountains. Rippling the
surface of countless lonely lakes.
And the rain, like a soft blanket, brings
its own perfect silence.*

Demand for Connemara marble flourished in the Victorian era. After the death of Prince Albert, Queen Victoria chose to stay in Britain, but her son Edward, the Prince of Wales, opened up a new era: the idea of royal visits overseas. As a young man, Edward travelled throughout southern Europe and the Middle East and visited India in the 1870s. Indeed, he introduced a new progressive style in the conduct of foreign affairs, forging closer ties between nations and strengthening British influence abroad.

After the death of Queen Victoria in 1901, Edward took the throne as King Edward VII. He had previously visited Ireland as Prince of Wales in 1868, and also in 1885, but that occasion was overshadowed by the land wars. Then, in 1903, he decided to make another visit – a royal tour taking in various locations including Derry, Dublin, Kerry, Cork, Galway and Connemara. Edward had gained great popularity as King and his skills in foreign affairs led him to be known as 'Edward the peacemaker'. A few months before the visit to Ireland, he had met with Pope Leo XIII, and this had won him great favour with the Catholic population. The King, accompanied by his wife, Queen Alexandra, together with the Royal entourage, arrived in at Killary Harbour, Connemara, on board the Royal Yacht, (the *Victoria and Albert*) on the morning of 29 July 1903, and the historic visit to Connemara took place. According to the *Illustrated London Times*, the visit was recorded as follows:

Their Majesties landed and made a tour of the Connemara district, visiting cottage and cabin, conversing in the most friendly manner with the natives. Their majesties travelled in a motor car. The following day the Connemara trip was brought to a close by a visit to Galway, where the King and Queen had an enthusiastic reception from a district that has not hitherto been deeply attached to the Crown.

A more detailed, and interesting account of the events is recorded by the *Galway Express* in their special supplement covering the Royal tour:

Their Majesties disembarked at Bundorragha at 10.45am, and proceeded by motor car to Recess, via Glenagiula, Leenane, Tully, Letterfrack, Kylemore and Lough Inagh. Recess was reached at 1.30pm.

A guard of honour of the Royal Irish Constabulary, commanded by the County Inspector, was stationed at the landing stage at Bundorragha.

Luncheon was served at Recess.

After luncheon, the King and Queen departed to inspect the Marble Quarries in the vicinity and afterwards departed by a special train to Galway, arriving at 5.15pm.

A further supplement in the *Galway Express* covers the visit in more detail:

Journey to Recess

The route led them first through Tully where a large crowd was assembled. An arch spanned the road bearing the words 'A friend of the Pope' and
at each side a Roman Catholic clergyman stood waving a red flag, whilst the
crowd acclaimed their Majesties to the echo of the same greeting.

The Graphic (a weekly illustrated newspaper of the day) carried a magnificent
illustration of the Royal visit to the Lissoughter marble quarries. The story that
accompanies the picture describes the scene somewhat charmingly thus:

Awaiting the arrival of the Royal car at Recess Railway Station was probably
the most extraordinary escort which had ever attended a King in Ireland.
It was composed of the Connemara Cavalry, tenant farmers and occupiers
of small holdings, the leaders bearing Union Jacks, and every rider wearing

The Royal party at the quarry.

streamers of red, white, blue and green. Their Connemara ponies were, as usual, excellent, but the harness was frequently nothing but rope, and the younger fellows rode bare-backed. After luncheon at the hotel a visit was paid to the marble quarries near Recess. To drive their Majesties thither, wrote the correspondent of the *Daily Telegraph*, an extraordinary equipage was brought to the hotel door. It was a landau in a most dilapidated condition. The handle of one of the doors had disappeared, the seats were wet from rain, and there was no brake. Attached to the carriage were two horses, which did not match any more than did their collars. At the back of the vehicle were stowed several yards of rope, designed to check its progress, if necessary. It was doubted whether their Majesties would enter such an equipage, but King Edward apparently recognised that great disappointment would result if the visit were abandoned, and resolved to carry out his programme. He entered the carriage without a smile at its singular appearance. Following His Majesty were the Queen, Princess Victoria, and the Lord-Lieutenant. In a second wagonette was Lord Knollys. On their way to the quarries the Royal visitors picked up an escort of the Connemara Cavalry. The entrance is by way of a soft and comparatively unmade road, running up a steep hill. The Royal equipage could make little progress, and eventually looked as if it would be stuck in the mud. Plenty of assistance was at once forthcoming. The people rushed forward to the carriages, nobody restraining them, and everyone anxious to help. In this way the Royal equipage was pushed forward, the King and the Queen being visibly delighted. Her Majesty, indeed, turned around every now and again and expressed her gratitude.

At the Quarry there was a large crowd assembled. Their Majesties watched with amused surprise the agility of the barefooted boys and girls in clambering from rock to rock. They examined with interest the working of the wire saws, which was explained to them by Mr Peter Rafferty and Mr St John Lyburn, who had the honour of being presented. At the conclusion of the inspection a VERY HANDSOME PRESENTATION was made to their Majesties It consisted of two beautiful specimens of Connemara marble, a harp and an inkstand magnificently mounted in silver and gold. The harp, which was a model of Brian Boru's harp in the museum, was presented to the Queen, and the inkstand to the King. They graciously accepted them and returned thanks.

Sadly, no trace can be found of the presentations made to the Royal couple, despite exhaustive research through the Royal Collection Trust at St James' Palace, London.

Another record of the visit from the book *Memories: Wise and Otherwise* by Sir Henry Robinson talks about the events of the famous day,

The Marble Quarries visited by the Royal party were up a mile of precipitous road at the back of the hotel. The stone is of the lovely green serpentine variety, which takes a beautiful polish and is used for anything from the

pillars of a church to a pair of cufflinks. The finest specimen, I believe, is the pillar in the Mutual Insurance office in New York. Mr Peter Rafferty, a returned American, was the lessee of the quarries, and hearing that the King and Queen were going to honour him with a visit, had two specially selected pieces of marble carved and polished as a silver mounted inkstand and a little Irish Harp, the inkstand being for the King and the harp for the Queen.

Among other things, Mr Rafferty told of an incident on the quarry during the visit that had made a tremendous impression on him:

'The Princess Victoria' [the King's daughter] he said 'in stepping out of the carriage had brushed her skirt against the wheel and had splashed it with yellow sand. And the Queen,' said Peter, 'the Queen of England, mind ye, stooped down as humble as the poorest woman in the land and brushed the

Accounts of this remarkable visit and anecdotes around it survive to this day in the folk memory of Connemara people.

Princess's skirt with her own hand! There she was with the First Lord of the Admiralty beside her, who she could have ordered to do it, and other great lords and generals, but no! Humble she was, and humbly she brushed the skirt with her own hand before anyone, and an example to everyone.'

Accounts of this remarkable visit and anecdotes around it survive to this day in the folk memory of Connemara people. It is indeed a remarkable story, especially when one stands in the wild and isolated landscape today, at Recess, for example, where the road turns up the hill to the quarry at Lissoughter. A greater quietness is hard to imagine than in this landscape, with its ancient mountains and lakes, serene and peaceful, and no disturbance other than birdsong, or the bleating of sheep, or the wind. And yet, one can also imagine the excitement of that celebration over a century ago, of the royal visit, and the spirited welcome in the local people as they came out to see the King.

Lifting The Marble From The Earth

Only a man of Connemara
senses where green veins are found.

On a damp Connemara day, with a fine misty rain sweeping across the fields, the land changes from moment to moment, and on the rocks and stones the colour comes to life. One can only imagine early man seeing the pure, translucent green marble glisten in this light and realising how it stands apart from common stone and is something unique.

So, how does one get the marble from the ground and transform it into a thing of beauty? There are several methods to achieve this goal. The earliest and simplest is referred to as 'picking' or 'scavenging', where the small loose stones are simply lifted by hand, or chipped away with hammers. This is an irregular method, and will only yield small blocks, but can often be of the highest grade of marble.

As the quarries opened in the nineteenth century, the stone was cleared of mud and debris by hand, and a method, first developed by the Greeks, was employed. This is known as 'line

drilling' or 'plugging and feathering'. This method calls for the drilling of vertical, closely-spaced holes along a line where a block is to be extracted from the quarry face. Steel wedges, or 'plugs' are put into each hole and two steel wedges, or 'feathers', placed on either side of the plug. By tapping these plugs, one by one, along the line of the holes, the pressure gradually builds up and the block will break evenly along the line of the holes. This can be a slow and painstaking operation, and if rushed, the block may break in an uneven fashion, and then all the drilling and preparation work will have been wasted.

The development of dynamite in the mid-nineteenth century provided a fast and simple method of quarrying that seemed like a quick and easy solution. Dynamite was used extensively and successfully for rock clearing for railway building, and for the extraction of rough stone for road making; little did the users of dynamite in Connemara realise the untold damage it would do.

Blasting was used on some quarries with disastrous effects. Although it removed many large blocks, the shock waves from the blasts not only cracked the extracted stone, but also went deeper into the quarry and opened many hair-line fractures in the unworked marble. Hence, when the marble was later extracted either by further blasting or other methods, the stone was filled with tiny fractures, and the marble cracked and broke in the production processes and in finished goods.

In the book *Practical Geology and Ancient Architecture of Ireland* (published in

1845) the author, architect George Wilkinson, warned of the dangers of the use of dynamite in marble quarries,

> The green coloured marble occurs in blocks, having a very rough and rugged surface, either loose and overlaying the surface near their natural beds, or in solid rocks, the extent of which is not however great. The appearance of the external surface contrasts very strongly with the sawn sections of the stone. In raising the marble, blasting should be avoided as much as possible, for it shakes the rock, and not only injures its appearance when worked, by causing cracks or small fissures, but frequently prevents its being converted to the purposes required, and for which it would otherwise be adapted. The frequent use of gunpowder in quarrying what has been hitherto raised, has very much prejudiced the sale of the marble, in consequence of the injury done to the blocks.

We owe our thanks to the Italians for the development of the wire saw in the late nineteenth century. They discovered that it was possible to cut marble by the use of a flexible steel wire, formed in a continuous loop, cutting vertically or horizontally through the rock face. A quartz sand, which is very abrasive, is dripped onto the wire to give it cutting 'teeth' and by the clever use of pulleys, a clean and regular cutting edge can be obtained. This method was employed in the Lissoughter quarry in the nineteenth century, and the wire was powered by a large steam engine with a huge fly-wheel on its side, through which the wire passed. A series of pylons and pulley wheels ensured that the wire could be actually cutting several faces at the same time. This method led to the extraction of many fine blocks of marble, and an ability to produce larger slabs with minimal damage or impact caused to the unquarried marble. In latter years with the development of synthetic diamonds, the steel wire is now fitted with diamond 'beads' which give a sharper, faster cut. Nowadays the driving machines are powered by electricity, allowing easier mobility on the quarry site.

One can only imagine early man

seeing the pure, translucent green marble

glisten in this light and realising how it

stands apart from common stone, and is

something unique.

Cutting and Polishing Marble

For every stone there is a moment
of knowledge.
Diamond-tipped steel penetrates
the hard casing.
And the stone reveals its secrets.

Once a rough block of marble is lifted from the ground, a process of transformation begins. Whether the piece is destined to be large or small – a huge wall panel or a tiny item of jewellery – the processes are similar. Using simple tools and age-old skills the rock is transformed – almost magically – from rough stone to polished beauty.

Firstly, the extracted block is assessed for colour, dimension and any natural flaws or fractures. An experienced stone worker will strike the block with a heavy hammer and if a clear 'ringing' sound resonates then the block is more likely to be free of any inherent cracks or flaws. For large pieces for use, say, in wall panels, the first cutting (or 'primary sawing') will then take place. The block is placed in a frame saw, and a multi-bladed crosscut sawing of the block takes place, using diamond-tipped blades. The operation takes several hours to complete and yields the original block cut into many large, thin slabs, about an inch, or 30mm, in thickness. These are then carefully removed and polished using a 'Jenny Lind' polishing system, whereby abrasives of various grades of coarseness will erase saw marks and scratches. Water is constantly washed over the marble during this procedure.

For smaller production (for example, jewellery pieces) the primary sawing is made using a large circular saw with diamond teeth. This block will be cut into very large slices of about 4 inches or 100mm in thickness. These are then inspected so only the most beautiful and patterned marble goes on for further processing. The slabs are cut again into smaller blocks about the size of a loaf of bread using the same circular saw technology. Once again, the marble is screened so that only the best is used. It is fair to say that, given the necessity of selecting only the finest marble, only about 10 per cent of quarried marble will be used in final production, the other 90 per cent being rejected or cut away in the production process.

In the case of jewellery, age-old craftsmanship is still employed. Skills developed in ancient Greek and Roman times are used to this day to achieve the complex geometry required to produce stones of beauty and symmetry. For most stones used in jewellery, the marble is cut into small rectangles, about 20

Skills developed in ancient Greek and Roman times are used to this day to achieve the complex geometry required to produce stones of beauty and symmetry.

Working the marble at the lathe.

per cent larger than the finished stone required. Each stone is then glued to a 'dop stick' using a heated wax. These sticks were originally held in the hand of the craftsman and the stone was gradually sanded down to the required oval or round shape; the domed effect was achieved by evenly grinding down the outer edges. Great skill and precision were required in order to ensure that the stone would fit accurately into the piece in which it was to be set. German technology, developed in the nineteenth century, has semi-automated this process: the dop stick is now made from brass and is fitted into a stonecutting machine, which will follow a pre-set pattern to ensure that the correct dimensions and an even, domed finish are achieved. The polishing process then begins. Just like the large slabs – but on a miniature scale – the stone is held against abrasives to eliminate saw and scratch marks. As the fineness of the grit increases, the surface becomes increasingly smooth. The secret of the sheen of Connemara marble lies in this perfectly-honed surface, which allows the stone to act as a mirror – drawing in light and reflecting back its own green deepness.

The Major Quarries Of Connemara

*Mountains split by bare hands and
tools of centuries past.
The stone holds traces of the rivers, the
mosses, the clays and the shapes of
ancient Connemara.*

As you drive along the road to Clifden you will notice, on the northerly side, places where the ground is more fertile than usual in the wild landscape. Typical Connemara land is poor and boggy, but the rich limestone seam of marble hidden underneath the ground enhances the quality of the soil and promotes better growth. Cultivated fields of a rich green provide a hint that marble can lie beneath. In several places the marble actually lies on the surfaces, and it was from these places that man could see this beautiful and magical stone, and realise it was something special.

Only three locations have provided enough accessible marble to be workable, and they are Lissoughter, Barnanoraun, and Streamstown; but in several other places, outcrops of marble can be seen. These lie on lakeshores and in remote

hillsides, and some have been explored but never opened fully. Who can guess what buried treasures still lie undiscovered beneath the Connemara hills?

BARNANORAUN

Barnanoraun or *Barr na nÓran*, takes its name from the Irish 'top of the spring wells'. Located in the Owenglin Valley, it was part of the Martin estates, and it was here that Thomas Martin opened the quarry in the early nineteenth century. The Martins built Ballynahinch Castle in the 1790s when the family was at the height of its power, and tables and fireplaces in the castle were made from this marble. The marble from this quarry is a rich green and yellow shade, with very fine patterning. As the Martins owned both Barnanoraun and Lissoughter quarries, there can sometimes be confusion when detailing the exact origin of some of the marbles used in various buildings. However, a very fine example of the marble from Barnanoraun can be seen in the geology building at Trinity College, Dublin.

As a consequence of the famine, the Martin estate became bankrupt and was offered for sale in 1849. It failed to attract any offers and lay unsold for several years. However, in 1872, it was sold to Richard Berridge, a brewer from England. Ballynahinch Castle itself changed hands many times and is now run as a luxury hotel, famed for its beautiful rooms and excellent salmon fishing. The estates, including the two quarries, Barnanoraun and Lissoughter, remained in the hands of the Berridge family until the 1960s. Michael Joyce from Recess purchased both quarries from the Berridge family in the 1960s, and for many years he ran a marble workshop on the site of the former Recess hotel, and a large craft shop nearby. In recent times his son, Kevin Joyce, embarked on an ambitious project to re-invigorate Barnanoraun and develop it for the production of large-dimensional stone for the building industry at home and overseas. Vast amounts of spoil and waste workings had to be removed and tons of earth moved to expose the deepest parts of the quarry. Constant pumping was needed to keep the workings dry and wire saws were used to cut into the surface stone and expose new faces for working. A state-of-the-art workshop has been developed in Recess for the processing of slabs and tiles, and several prestigious projects have chosen this marble, including the reception area at Dáil Eireann (the Irish Parliament) in Dublin, the Galway Mayo Institute of Technology in Galway city, and many other important buildings and installations worldwide.

Opposite: Gleninagh and the Twelve Bens from Lissoughter hill.

STREAMSTOWN QUARRY

Two miles north of Clifden is Streamstown Bay, an inlet of the Atlantic. The name comes from the Irish, *Barr an tSrutha*, meaning 'the top of the stream', and the area was one of the most prosperous places in Connemara in the eighteenth century because of a lively trade in smuggling. This area contains evidence of the earliest settlers in the region, and many forts, ancient tombs and churches can be found. There is a small copper deposit in the area, as well as a significant marble deposit. The area was incorporated into the estates of the D'Arcy family,

Clifden Castle.

which they were granted when Gaelic clans had their lands confiscated in the Cromwellian Wars. After John D'Arcy inherited the estate in 1804 he set about developing the town and its assets. He built a fine castle overlooking Clifden Bay in 1815, and this remained the family seat until the estate became bankrupt in the 1840s. The estate was then sold to Thomas Eyre of the city of Bath, in England, who used it as a summer home, but abandoned it in 1894. Today, the empty shell of this majestic building can still be seen, the poignant remains of the once thriving estate.

The marble quarry at Streamstown was worked extensively in the nineteenth and twentieth centuries, and has two distinct colours in its marble – a rich patterned green and a brownish-white, known as 'sepia marble'. Good examples of the marble from Streamstown are the interior columns in the church of Saints Peter and Paul in Athlone and the floor in Galway Cathedral. The quarry at Streamstown remained open as other quarries fell into disuse, largely because the owners used innovative methods of quarrying and production to produce large-scale pieces of stone suitable for use in buildings. The owner of the quarry was Terence Bourke, the 10th Earl of Mayo, and his company, Marble Panels Ltd., produced many marble pieces for the home and export trade. In fact, almost all large pieces of Connemara marble from the mid-twentieth century came from this quarry, and this kept the 'brand' alive.

Terence Bourke used a *terrazzo* method to produce durable floor tiles, and these were used in the Dublin Airport Terminal built in the 1970s. He also utilised the technique of 'quadrilateral symmetry' to enhance and harmonise the complex patterns in the marble. This involved cutting a large block of marble into slices and fitting the cut slabs together in such a way that the design appears in a kaleidoscopic or butterfly type of pattern. Several fine examples of this can be seen in the Roche pharmaceutical facility in County Clare. Another innovation from Streamstown was the production of lightweight marble panels. This involved a lightweight honeycomb frame onto which very thin sheets of marble were attached using special adhesive. The lightness inherent in using a veneer of marble meant that the panels could be used in luxury jets, cruise liners and private yachts – where weight would otherwise have been an issue. Sadly, however, popularity and demand did not live up to expectations and this business ceased. Quarrying was abandoned there for some years, but in recent times it was taken over by Ambrose Joyce from Moycullen in Galway. This latter family operates the Connemara Marble Visitor Centre at Moycullen, and the marble now quarried at Streamstown goes to make the items on sale there.

LISSOUGHTER

The quarry at Lissoughter nestles into the hill just above the small settlement of Recess. The name 'Lissoughter' comes from the Irish, 'Lios Uachtair,' 'lios' refers to an enclosure – a dwelling or ancient ring fort, or even a fairy ring. It can also mean a halo or ring of light. The word 'uachtair' means 'the upper part of the mountain'. The distinctive conical hill into which the quarry is set is called Cnoc Lios Uachtair, but is known locally as 'Glass Mountain' because of the way its quartzite slopes reflect the sunlight. But it is the buried treasure at the foot of the mountain that gives this place a special meaning, for it is here that the finest Connemara marble can be found.

In the fields around the quarry at Lissoughter it is still possible to see the remains of the potato ridges where this staple was grown for generations. Above the quarry, a few stone cottages remain, now roofless and ruined, and it was here that quarry workers lived in the nineteenth century. One of the cottages served as a shop where goods were bought and sold, and there is talk that another was a 'shebeen' or illegal bar. Near the road leading to the quarry you can still see the shape of a larger stone building, known locally (with a certain humour) as 'marble hall', where the workers gathered long ago. On closer inspection it can be seen that the walls are, in fact, made from the unused small rough blocks from the quarry, so that the name is indeed suitable for the old assembly hall. Local lore tells that this hall stood out from other buildings in the village. Apart from being much larger than the usual cottages, the 'marble hall' had the exotic embellishment of a tin roof, while the other houses had only simple thatched roofs to keep out the harsh Connemara weather.

Lissoughter was located in the estate of the Martins of Ballynahinch, and it was they who opened it in the early nineteenth century. With the estate going bankrupt, Lissoughter suffered the same fate as Barnanoraun, and it was sold to the Berridge family in 1872. By the turn of the century, the Berridges had granted a lease to work both the Barnanoraun and Lissoughter quarries to a Mr Peter Rafferty, who was a returned American, and it was under his guidance that a great deal of fine marble was produced.

Lissoughter Quarry. c.1900.

Lissoughter had a huge geological advantage over the other quarries. The marble was above ground and was extracted without the need for extensive deep diggings and the attendant problems that that would present. In addition, as the quarry was located on the hillside, it was free from flooding – a problem that dogged the other locations.

A wire saw was installed at the quarry, and a criss-cross network of wires moving through a complex network of pylons, pulleys and wheels, could saw several different faces at the same time. All this was powered by a steam engine in the centre of the quarry.

The stone from Lissoughter was transported to Cloonisle Pier, some six miles away, and then by sea to Dublin for polishing. The arrival of the railways eased the problems of transportation to some degree, but generally, the stone went either westward – usually to Clifden – for onward transport by sea, or eastward to Galway for onward delivery by the emerging rail network.

The Railway Hotel at Recess was especially built to welcome tourists to the region and a special siding was built for access to the nearby Lissoughter quarry. A sledway was developed to bring the marble down a one-mile trackway from the quarry. The sleds were attached to a steel wire and pulley system, and pulled by horses. Sadly, no trace remains of the sledway, but the system is remembered still in the lore of the locality, and there is a story, which recalls an incident in which the wire snapped while a large block of marble was being transported down towards the rail siding. The block careered downwards and the horse was killed. Luckily, no human lives were lost.

Peter Rafferty continued to operate the quarry leases on behalf of the Berridge family into the twentieth century. After some years, Festus Joyce from Recess took over the leases as part payment for financing the quarries. In turn, Festus Joyce's son, Patrick Joyce, bought both Lissoughter and Barnanoraun from the Berridge family in the 1960s. Lissoughter was little used from this time and largely lay abandoned. In 1983 I purchased the little piece of history that is the Lissoughter quarry so that I would have a ready supply of marble for my business.

In the quarry itself, there is a fairy tree, a whitethorn; its branches distorted and gnarled by the fierce Atlantic winds. Legend has it that these fairy trees stand where fairies gather at night, and it is considered very unlucky to damage or destroy such a tree. There is a long-standing tradition of respect for the fairies in rural Ireland; they bring good luck and protection, but they can be mischievous too! When working the quarry, the fairies are accorded their due, and when things go wrong – when a machine breaks down, or a tool is misplaced, or a chain snaps – it is the fairies who are to blame. If the quarrying is stubborn and the stone is refusing to budge from the ground, for sure it's the fairies holding on to the marble for themselves. And of course, when a block of marble is lifted, a word of thanks goes to the fairies for giving up the marble. Although never proven, the fairies are at work in the quarry, and are treated with due dignity – and with a twinkle in the eye too.

In the quarry itself, there is a fairy tree, a whitethorn; its branches distorted and gnarled by the fierce Atlantic winds.

As you drive along the road to Clifden you will notice, on the northerly side, places where the ground is more fertile than usual in the wild landscape.

Connemara Marble in the Twentieth Century

Connemara. The soul of Ireland.

By the beginning of the twentieth century, Connemara marble was at the height of its fame. It had become of such note internationally that an isolated quarry was deemed worthy of a royal visit in 1903. In fact, the Irish state railway coach that brought King Edward VII back from Galway to Dublin after his visit west of the Shannon had been specially constructed for the Great Southern and Western Railway at the Inchicore works in Dublin. It was decorated with panels of Connemara marble, and even the hand basins were carved from the finest serpentine stone.

Further fine examples of Connemara marble were installed at the newly-built Westminster Cathedral in London, which opened in 1908, and the nearby General Post Office (in 1910); the Museum of Antiquities in Dublin (in 1911). The new Pennsylvania State Capitol building (1906) used Connemara marble in its Senate Chamber. Even James Joyce's *Ulysses*, published in 1922 makes mention of a clock made from Connemara marble.

The onset of the First World War saw a winding down in building along Classical lines, and a new age in construction emerged in the twentieth century with different design techniques using more modern materials such as concrete, glass and steel. By the 1920s it seemed that the demand for Connemara marble was slowing down. When an architect called for decorative stone, these were now sourced from Greece, Italy and Africa as access to these locations had improved. There remained a limited, but steady demand for Connemara marble in the USA and Canada, mainly for church decoration. The Basilica of the National Shrine of the Immaculate Conception in Washington DC was one such project. This immense building began in the 1920s, but the project came to a halt in 1929 due to the onset of the Great Depression and did not re-start for several years. Connemara marble is most notably used at the Basilica in the chapel of Our Lady Queen of Ireland, where the walls are made of Connemara marble.

In tandem with the worldwide recession after 1929 and into the 1930s, the Irish economy too, was in a poor state. There were few industries and a heavy reliance on a largely agrarian economy. The famine of the 1840s had far reaching consequences attested to by the declining population and continuing emigration. Upheavals associated with the emergence of the Free State in the 1920s also contributed to economic instability. Demand for building and building materials was low and indeed, the only significant area where activity continued was church building. Surprisingly, church building flourished to some degree in this period, and there are many fine examples of Connemara marble columns, altars and decorative panels still extant all around the country from this time. That the industry struggled on can be seen in a report from the *Connaught Tribune* in the 1930s, which tells of a consignment of ten tons of marble from the Streamstown quarry being exported to Belgium, and that the new Saint Anne's Cathedral in Belfast had also taken delivery of a large quantity of Connemara marble.

In the 1930s and 40s several attempts were made by Government and State Development

There is an excellent example of the twentieth-century use of Connemara marble in a silver box with marble inlays, made in 1910 (above).

bodies to revive the industry and exploit the potential of the marble, but to no avail. The impending prospect of war in Europe coupled with the overall decline in economic activity and Ireland's protectionism stance all contributed to a dwindling of production within the marble business. Another blow to the region came with the closure of the railway linking Galway to Clifden in 1935.

In the 1950s and 60s there were several instances of Connemara marble being used as an external cladding on new buildings constructed at this time. However, the polished surfaces of the marble were found to be unsuitable for exterior use as they weathered badly and discoloured after a short period of time, leading to the abandonment of this ill-tested architectural experiment. Nonetheless there are still some interesting examples of the use of the marble to be seen in Irish twentieth-century interiors, including the entrance portico to the Department of Industry and Commerce Building in Kildare Street in Dublin (completed in 1942 and constructed during a period of war and a materials shortage); parts of the staircase in Áras Mhic Dhiarmada – the land-mark 'Busáras Building', Dublin's main bus station – near the Custom House in Dublin, the floor of Galway Cathedral, and the 1960s restoration of the floor in the State Apartments at Dublin Castle, which was exclusively made from Connemara marble.

Connemara marble jewellery at this time was largely made in Britain, as Irish jewellery tended mainly towards the ecclesiastical and table decoration market. The main skills in Irish jewellery were in the manufacture of larger pieces of silver, there being a fine tradition of antique Dublin silver – teapots, salvers and flatware. A handful of small craft shops produced the smaller decorative pieces for the jewellery market. It was not until the 1930s that larger-scale jewellery production began with the founding of O'Connor Jewellery works in Harold's Cross, Dublin, and the Irish Souvenir and Jewellery Company in Parnell Street, Dublin. These two companies were at the forefront of the emerging home-produced jewellery industry. They produced many fine designs fabricated in gold and silver, featuring Ireland's traditional motifs – the shamrock, the claddagh, Celtic crosses and Celtic knotwork, many of which were adorned with inlay of cabochons of green serpentine marble. There is an excellent example of the twentieth-century use of Connemara marble in a silver box with marble inlays, made in 1910, on display in the Museum of Decorative Arts in Dublin; and a very fine silver cup with a Connemara marble base made in 1949, on display in the Victoria and Albert Museum in London. The ecclesiastical demand continued too, and a specially-commissioned pair of Connemara marble candlesticks from the quarry at Streamstown was presented to Pope John Paul II in 1981. The historic visit of President John F. Kennedy to Ireland in June 1963 was marked by a number of presentations to him. These included silver caskets containing the Freedom of the cities of Galway and Cork, a replica of the Great Mace of Galway and a plaque bearing the Leinster Harp. All of the items included Connemara marble in their composition, and they are now placed in the JFK Memorial Library in Boston.

It is important to note the role of the assay office in the production of Irish jewellery; all articles made of precious metals must go to the assay office for testing before they can be released for sale to the public. Since 1637, each piece of jewellery is tested to verify that it conforms to one the precious metal standards as prescribed by Irish law, and when passed, receives the appropriate hallmark. The Irish hallmark is regarded as one of the earliest forms

Gifts presented to President Kennedy:
Leinster Harp Plaque, Freedom of the City of Galway
Casket, Freedom of the City of Cork Casket.

of consumer protection and gives the purchaser the guarantee that the jewellery they have purchased from Ireland is as described. The hallmark also incorporates the maker's initials and the mark of the Dublin assay office.

The quarries continued on, sometimes erratically, through the economically difficult times in the twentieth century. The innovative work by Terence Bourke, Earl of Mayo, at Streamstown quarry has already been mentioned, and his new approach to the production of Connemara marble by creating lightweight decorative panels was indeed very important to the industry. Investment at Streamstown in modern wire sawing machinery and the use of imaginative solutions (particularly quadrilateral and bilateral symmetry) to meet the demand in a changing building industry helped to sustain the idea of Connemara marble as an exceptional interior adornment in changing times.

At Lissoughter, there was a huge problem of 'overburden' in the quarry. As marble was being extracted, the soil and loose material had to be removed to expose the quarry face. The use of fairly basic lifting equipment meant that this material was deposited close to the quarry workings. After many years of this practice, the waste tips had become enormous and now covered areas in such a way as to prevent further extraction. It was estimated that these miniature mountains contained over ten thousand tons of debris. The expense that would be incurred moving this material would be huge and with a slowdown in demand, the operation of the quarry was becoming unsustainable. As a result the quarry slowly slipped into a dormant state, leaving moss and lichens to cover the exposed workings, swallowing the quarry back into the hill. The marble was to remain like a Sleeping Beauty, waiting to be woken once more. Local knowledge tells that the quarry lay idle thus until 1963, when the Berridge family sold the property to Michael Joyce from Recess. Michael made a few exploratory visits to the quarry, but never made further progress as his main interest was in his other quarry at Barnanoraun.

The quarry in Barnanoraun produced small slabs for the limited tourism market, which was still in its infancy. However, when ownership of this quarry passed to Kevin Joyce, there was significant investment in the exploration and reopening of this ancient location. Vast tonnage of loose rock was removed; extensive pumping equipment installed and a new quarry face exposed, revealing new deposits of the famous marble. No doubt the Martins would have been pleased to see life being breathed into these new workings. It was now possible to extract large blocks of this excellent marble and to process them for installation in the new buildings that were springing up at the time of Ireland's economic recovery at the end of the twentieth century. Notable installations from this period include the reception area at Dáil Éireann (the Irish Parliament), the Radisson Hotel in Dublin as well as the Galway Mayo Institute of Technology.

Connemara marble proved itself to be a survivor through the vicissitudes of the twentieth century. It managed to maintain a place in the building industry and sustained its centrality as a cherished souvenir item as well as a desirable stone for use in craft-work and jewellery. Its identity as a particularly Irish stone became more defined especially after the foundation of the independent Irish State, when the symbolism of Irish motifs was imbued with a greater intensity and sense of belonging. By the end of the century, Connemara marble was well poised to participate in the drive for exports as Irish manufacturers looked increasingly to markets overseas, and to play a new and yet again updated role in the global market.

Reopening of the Lissoughter Quarry

Hands to work the green marble out of the raw stone.
Fathers' and sons' hands that trace the veins and natural folds.
The passionate hands of a sculptor.

For me personally, this is a special story, and the beginning of my love for Connemara. I have spent my working career in a family business that produces a range of jewellery, gifts, religious items and souvenirs. Founded by my grandfather in 1945, the company was originally involved in the manufacture of Irish rosary beads, used for keeping count of prayers in the Catholic tradition. These were made originally from the horns of animals – and unique to Ireland – but as the supply of this raw material dwindled, another material was sought.

Connemara marble was an obvious choice, so under the

The Fairy Tree at Lissoughter.

This was Máirtín Burke (above with Stephen Walsh), who had lived in Lissoughter all his life, and who was a fount of local lore and knowledge.

direction of my father, the company set about developing the special Conne-mara marble double-link rosary, which was exported worldwide. Changes in Church practices in the 1960s resulted in a slowdown in the demand for rosa-ries, but by a happy coincidence, the fledgling tourism market was growing, and visitors to Ireland sought out indigenous Irish items featuring Irish designs and Irish materials. With knowledge of rosary making – beads and wire work – it was a logical progression for the business to develop jewellery and a range of other souvenir items made from the marble.

The company sourced the raw marble from Streamstown and Barnanoraun quarries, but supply was intermittent and as the company's business grew, my father and I agreed that we should try to ensure that a supply of the raw mate-rial on which our business was becoming increasingly dependent was readily available.

In the autumn of 1982, I made contact with the Geological Survey of Ireland – a government-funded body with expertise and detailed databases relating to all matters geological in Ireland. There I met some extremely helpful people who supplied me with historical references on quarry activities, a report detail-ing a study of the Connemara marble industry, and some old maps of marble locations. So, armed with this information, I set off west to search for Conne-mara marble. My journey brought me through the marble-bearing region of County Galway, from windswept mountaintops to deserted valleys. In some cases, I would find a marble outcrop, but the location would be so remote as to make it utterly inaccessible for extraction purposes. In other cases where the marble was accessible, the quality of the stone was inad-equate, as it was too pale or flaky.

After miles and miles of driving, hunting and questioning, I came to the little village of Lissoughter. On the map, the caption read 'abandoned marble quarry', and this aroused my curiosity. At a bend in the road I could see an opening into the hillside, completely overgrown with briars and brambles. Suspecting this could be the entrance, I parked the car and proceeded to scramble and hack my way through the mess of thorns. There was a clearing beyond, completely deserted but for the sound of the wind and the distant bleating of sheep. From the side of the slope I could see a mossy greenish-black stone protruding. I chipped away at this stone and beneath the lichens and dirt I could see green in the very stone itself – this was indeed the old quarry that had lain dormant for over sixty years.

Retracing my steps to the car, I saw an elderly gentleman watching from his cottage. I approached and introduced myself, and the gentleman confirmed that it was the old quarry of Lissoughter and that he himself had worked there, as had his father before him. This was Máirtín Burke, who had lived in Lissough-ter all his life, and who was a fount of local lore and knowledge. Máirtín told me that the property was owned by a Michael Joyce from the nearby village of Recess, and I immediately set out to meet him. Fortunately for me, Michael was no longer interested in the property, as his main interest lay in his other quarry in Barnanoraun, and he was prepared to sell, so by the end of a memorable day, I was the proud owner of a quarry in Connemara.

A subsequent survey by an expert geologist confirmed that there were good reserves of stone lying beneath the surface, and work to clear the debris and reopen the quarry could begin. Tons and tons of the overburden (the old waste heaps) were removed to expose the old quarry faces, and various artefacts from the old times were recovered – including pulley wheels, bolts and anchors from the old workings, quarry chains and even a clay pipe. Máirtín Burke and his son Marty told stories about the old workings, and Máirtín (in his eighties by then) recalled the days when he worked extracting the marble, and pointed out the parts of the quarry that yielded 'the finest marble in the world'.

What fascinated me was the pace of life in Connemara, and how nature's time ruled over the clock. I recall arriving at the quarry one morning to super-vise work only to find the place abandoned, and not a soul in sight. It turned out that sheep had to be rounded up from the hillside ready for market that

Kevin Andrews and Stephen Walsh exploring the old quarry face in 1984.

day, and this task was to be completed before quarrying could begin. As Lissoughter is located within the Gaeltacht (the area where the Irish language is predominantly spoken) the locals would chat and banter in Irish in the lilting Connemara dialect, while I would often not grasp what was going on – due to my meagre school-level knowledge of the language. It was, however, wonderful for me to hear the language used in an everyday context with such mellifluous fluency.

Life in the west also had a sort of time-lag effect if things went wrong. In the pre-mobile phone era, a trip to the local post office in Recess would have to be made to get word to a supplier for a spare part, which was often the case if a hydraulic hose broke or a chain snapped. If the part had to come from Dublin, as it so often did, it could take a day to reach Galway, and another to reach Connemara. As an eager young man, the thoughts of standing idle for a whole forty-eight hours seemed a lifetime, but the locals would take this in their stride and divert their attention to another task – making hay, tending sheep or cutting turf from the local bog.

The weather, too, played its part, as rain could sometimes stop operations when the ground would become too muddy to operate machinery in safety. Work would stop until the clouds parted. But, by contrast, if the weather was fine, work would not cease until the last block would be lifted, or the last truck despatched, often well into the long, late, bright Connemara evening.

Time spent on the quarry has taught me a great deal. An understanding of Connemara people, their inimitable capacity for life at a different pace, their connection to weather and remoteness, and their sense of pride in and absolute belonging to the magnificent place they call home. It has been my privilege to be connected to this, to have been able to sit and take tea and listen to stories, unhurriedly told, of current and bygone days alike.

Watching the marble come from the earth where it has been secreted for millions of years was a most memorable and exciting experience – to be the first to feast my eyes on the unique colouration and patterning that had formed aeons ago. And it was remarkable to see majestic nature come to life in the hands of the quarryman, and then to be crafted into beautiful finished artifacts.

I could now understand how ancient man revered these unique green marbles, and I could connect to the excitement that must have prevailed when the D'Arcys and the Martins began to uncover the deposits and watch the first marbles being lifted and used with pride, in fine buildings and furnishings.

Connemara
Marble Today

It is said that no single piece is the same,
each having its own evolution.
The same colour, same pattern will never
be seen again.

I have traced the story of Connemara marble from its geological formation, through the uses made of it in prehistoric times, to the opening of the quarries, and the uses to which it was put in buildings, furniture, souvenirs and jewellery.

Connemara has an interesting history. It is humbling to think of early settlers forging a life on the rugged shores, and of the adventurers of later generations who sought a home in this remote place. Much has been written about the great families, the hardship of famine and social and political events that occurred over the centuries, but the marble lying beneath the surface upon which history was formed, was overlooked. In fact, Connemara marble has generally been no more than a footnote to history. But it has an illustrious history of its own, and a story around it that is worth the telling.

There was a tendency for all things Irish to be undervalued here at home in the 1960s and 70s. Ireland was turning outwards to the wider world, particularly after joining the European Economic Community in 1973. There was a yearning for international brands and a more 'European' look in fashion and design. There was, too, a reluctance among the Irish to embrace the age-old iconic symbols and designs of their own heritage. Perhaps it was a lack of confidence in Irish-produced goods when confronted with the vast array of products available from Europe. Or perhaps the reluctance to embrace the history and heritage of Ireland was due, in part, to the political tensions of the late 1960s and 70s. There was prevailing notion at the time that it was essential to bury the past in order to avoid conflict.

The Industrial Development Authority (IDA) encouraged overseas investment in Ireland, and many overseas companies set up operations here. As Irish industry became more technically advanced, and the Irish consumer more interested in imports, the traditional crafts sector got overlooked and there was less emphasis placed on the development and preservation of many national traditional skills.

At the same time, the increased availability of air travel brought a new influx of visitors to Ireland, and the souvenir business responded. Sweaters, shawls, jewellery, crystal, pottery and gifts were suddenly in demand. In the absence of design-led production and a lack of investment, some producers fell into the trap of making products that often fell well below the expectations of the sophisticated traveller. Journalists who wrote of 'knick-knacks' and 'gaudy shamrocks' undoubtedly had some justification for their criticisms of poor standards and kitsch in the souvenir market.

However, Enterprise Ireland set about encouraging new fashion and jewellery design, while companies like Waterford Crystal, fashion designers like Sybil Connolly, organisations like Kilkenny Design and individuals like Louis Mulcahy were tireless ambassadors for quality Irish giftware. From all of this grew a better understanding of the importance of excellence in Irish design, and the idea that products with a contemporary look could simultaneously hold on to an inherent 'Irishness'.

An experience I had in the 1980s illustrates the typical double-think in relation to Connemara marble. Having developed a high-quality range of Conne-

1980s: Connemara Marble desk stationery.

mara marble desktop accessories, I went to the stationery department in one of Dublin's leading department stores. The buyer took one look at the products and immediately rejected them, saying, 'These are made from Connemara marble, that's only for the tourists, please make an appointment to meet the souvenir department.'

Having made a fresh appointment to meet the souvenir buyer in the same store, I was met with the remark, 'These are really beautiful and would be best suited in the stationery department, as they are definitely luxury gifts.'

I was now – appropriately enough – caught between a rock and a hard

Above and following pages:
Antolini projects using
Connemara marble.

place! This just about summed up the attitude to Connemara marble in the domestic market at that time.

Economic growth and an easing of political tensions in Ireland in the latter years of the twentieth century brought about a renewal of pride in being Irish. The new air of confidence carried through, breathing fresh life into old designs and old materials. The claddagh, the shamrock, cable knitwear were all reinvigorated by better design, and there was a deeper understanding – and acceptance – of their significance and symbolism.

The quarry at Barnanoraun, under the guidance of Kevin Joyce, was now exporting blocks of marble to the Italian company, Antolini, who specialise in rare and luxurious stone for installation in unique and prestigious projects. This was a real endorsement of the value of green Connemara among the world's rarest and most beautiful marbles, and it is now being placed in prestigious projects around the globe.

New higher-quality products caught the attention of the international market, and companies such as Bloomingdale's, National Geographic Society, QVC, Duty Free Shops, specialist retailers and many independent jewellers recognised the quality and beauty of Connemara marble.

Design in marble jewellery now harmonises

the traditional material with a contemporary look of high-quality manufacture. The discerning consumer may value a piece of Connemara marble jewellery because of its connection to Ireland, or they may wish to own it simply because of its innate stand-alone beauty.

Connemara marble is as intrinsic to the landscape as the flora and fauna, the bogs, the trees; its form and beauty evoke everything that is beautiful and cherished in Ireland – the wild landscape, the sea-green of seascape, the swirl of wind and the depth of time. The stone holds its beauty like an ancient secret that has drawn people from time immemorial, and that now draws us, so that again and again throughout the generations, we are captured by its spell.

Design in marble now harmonises the traditional material with a contemporary look of high-quality manufacture.

QVC and Connemara Marble – A Personal Adventure

A rare green stone
scattered across the world
to its people.

In 1989 an unexpected turn of events was to bring Connemara marble to a wider audience.

That year, I was approached by Enterprise Ireland – the State organisation designed to promote Irish export sales. They had received an enquiry from a newly-formed American business interested in selling Irish items on television. A team of buyers was coming to Ireland to meet with companies interested in this new form of business.

The company was QVC, formed a few years earlier in 1986 by Joseph Segel in West Chester, Pennsylvania. Their message was simple – a commitment to Quality, Value and Convenience. QVC rapidly gained consumer confidence, brand loyalty and a strong position in the market place. Customers were offered a wide range of new and interesting products, which they could access from

the comfort of their own homes. The hosts explained the key features and benefits of the products for sale, and demonstrated gadgets and devices, thus bringing the products to life for the viewer.

I thought this was the craziest idea ever! The idea of selling sweaters, jewellery and even fragrances on television seemed weird and I thought the idea would never succeed. On the day the QVC buying team arrived, I was busy, so I sent a junior employee to the meeting on my behalf. He reported that the concept was indeed strange, but out of curiosity, as well as courtesy to the visitors, he dutifully followed up. It seemed that Connemara marble had impressed the buyers; in particular, an item known as '*An Cloc Cosánta*' (meaning 'the Luck Stone') – a simple necklet of Connemara marble on a cord. After overcoming my initial scepticism, we accepted an order for this product.

St Patrick's Day, 1990, arrived and the first ever Irish show – a three-hour special – took place. It was a huge success; everything sold out including my luck stone. The *Cloc Cosánta* had really worked! This was the beginning of the St Patrick's Day Special – an event that has now become a veritable tradition, a platform for Irish vendors to connect with the American marketplace through a particularly modern medium. Of course, the ancestral and family ties of many Americans to Ireland are evident in the strong desire for Irish-made items, particularly those resonant of Irish culture and tradition. Connemara marble has featured on QVC on St Patrick's Day and at other times over the years, in many forms from rosary beads

Above: Stephen Walsh, Jane Treacy and John Cullen.
Opposite: the famous double-link rosary.

to jewellery. The American connection has become another interesting dimension in the Connemara marble story.

In 1991, the buying team again travelled to Ireland. Two presenters and a camera crew came to take some scenic shots for the broadcast. It was suggested that the quarry would be a good location for filming, and so it was that I met with the hosts Pat James-DeMentri and Jane Treacy. On the appointed day, I was met at the quarry gates by the QVC team, and the cameramen set up the shot so that Pat could talk to camera. While this was happening, I got chatting to the other host, Jane Treacy, and was immediately captivated by her charm, her passion for Irish heritage and love of things Irish. We immediately seemed to click and a wonderful trans-Atlantic friendship had begun. After filming, the entire crew headed for the Abbeyglen Hotel in Clifden, famous for its fine food and hospitality, and after dinner the owner, Paul Hughes led us all in a sing-song and dancing. This made a great impression on the American visitors who were overwhelmed by the warmth of Irish hospitality and the great sense of fun.

In 1993, I was invited to appear on air on QVC on St Patrick's Day. The thought of appearing live on television, beamed into millions and millions of American homes struck sheer terror in my heart!

To my great relief, however, the show was hosted that day by Jane Treacy, who I had not seen since the sing-song in Clifden, and she immediately put me at my ease. I felt sure she would bring me the luck of the Irish! As the show went on air, my knees were knocking in fright … and then we were live. Jane introduced me and asked me, 'Stephen, what is that beautiful green stone and what age is it?'

As I answered 'Connemara marble … and 900 million years old', I at once felt at home with the familiarity of my subject. We went on to chat about the Connemara marble rosary that was being offered, and in no time at all, Jane was bringing our conversation to a close as they had all sold out. My television debut was as friendly and relaxed as the first afternoon we had met on the side of the quarry in Ireland.

Over the past twenty-five years I have worked with many wonderful QVC hosts, and I have enjoyed the tremendous camaraderie among all the Irish vendors who travel to QVC from Ireland. Above all we have all been privileged to make a connection with an audience appreciative of things Irish.

Some have no Irish connections at all, and enjoy the products simply for what they are. Others have Irish ancestry and a keen sense of their Irish heritage; their desire to connect with Ireland is palpable and reminds us all of the history that scattered the Irish worldwide, as well as the things that unite us – a sense of place and culture held deeply by Irish people every-where, united always by character and humour and a connection to home.

Opposite: Jane Treacy and Stephen Walsh appear on QVC in 1993. Above: the Walsh brothers filming for QVC.

STEPHEN WALSH

STEPHEN WALSH was born in Dublin and educated by the Carmelite Fathers and the College of Commerce. Stephen has spent all of his working life in the family-run business, J.C. Walsh & Sons Ltd., which is now in its third generation.

As Managing Director, Stephen has led the company into new and varied markets, and has gained recognition in the business community, having received awards for design, innovation and enterprise. He has a keen interest in jewellery design and attends the leading jewellery trade fairs worldwide.

Married with three adult children, Stephen has a love of outdoor pursuits and enjoys hill walking and long-distance running. He is a former president of the Dublin Junior Chamber, board member of the Dublin Chamber of Commerce and is currently an active member of the local business network.

CONNEMARA MARBLE

*'There is a quiet corner of my soul that sits apart,
And though the world may tumult elsewhere
This holds peace.'*

*'Light never so green, so blue falling
over the mountains. Rippling the
surface of countless lonely lakes.
And the rain, like a soft blanket, brings
its own perfect silence.'*

'Connemara. The soul of Ireland.'

*'There are whispers of history in the wind.
Mist smelled in the air, in the crumbling
dense black turf.
And precious stone buried in the mountains,
Ireland's national gem.'*

*'Mountains split by bare hands and
tools of centuries past.
The stone holds traces of the rivers, the
mosses, the clays and the shapes of
ancient Connemara.'*

*'The search for pure, translucent
green – the vein of serpentine
flowing through the stone.'*

'The rarest Connemara Marble of all.'

*'Marble twisted and shaped before
timebegan.Form and beauty
buried in simple stone.'*

*'Only a man of Connemara
senses where green veins are found.'*

*'For every stone there is a moment of
knowledge.
Diamond tipped steel penetrates the hard
casing.
And the stone reveals its secrets.'*

*'Hands to work the green marble out of
the raw stone.
Fathers' and sons' hands that trace the
veins and natural folds.
The passionate hands of a sculptor.'*

'It is said that no single piece is the same,
each having its own evolution.
The same colour, same pattern will never
be seen again.'

'Nine hundred million years the marble lay
beneath the mountains of Connemara.
Until craftsmen shaped it for those of us
who have a passion for beauty.
For the past.
For the wild and lonely soul of Ireland.'

'One day the rocks will give up their last vein of marble.
And who will hold the precious stone?'

BIBLIOGRAPHY

Dolkart, Andrew S., *Morningside Heights, A History of its Architecture & Development* (New York, 1998).

Gibbons, Michael, and Gahan, Reingard, *Connemara, Visions Of Iar Connacht* (Northern Ireland, 2004).

Harbison, Peter, *Pilgrimage in Ireland, The Monuments and the People* (London, 1991).

Lynam, Shevawn, *Humanity Dick, A Biography of Richard Martin, M.P. 1754–1834* (London, 1975).

Max, Michael D., *Connemara Marble And The Industry Based Upon It, Geological Survey of Ireland, Report Series 85/2* (Dublin, 1985).

O'Connor, Kevin, *Ironing The Land, The Coming of The Railways to Ireland* (Dublin, 1999).

Ó Riordáin, Sean P., *et al*, 'Lough Gur Excavations, Further Neolithic and Beaker Habitation on Knockadoon', *Proceedings of the Royal Irish Academy, Vol 87C* (1987) pp. 299 – 506.

Pine, John, *Connemara Marble Occurrences, Geological Survey Of Ireland* (Dublin, 1980).

Ridley, Jane, *Bertie: A Life Of Edward VII* (London, 2012).

Robinson, Tim, *Listening to the Wind* (Dublin, 2006).

Robinson, Tim, *The Last Pool of Darkness* (Dublin, 2008).

Robinson, Tim, *Connemara, A Little Gaelic Kingdom* (Dublin, 2011).

Rogers, Patrick, *The Beauty Of Stone, The Westminster Cathedral Marbles* (London, 2008).

Villiers-Tuthill, Kathleen, *A Colony of Strangers, The Founding and Early History of Clifden* (Clifden, 2012).

Whilde, Tony, *The Natural History of Connemara* (London, 1994).